D1490992

Understanding
Boat Batteries
and
Battery Charging

Understanding
Boat Batteries
and
Battery Charging

JOHN C. PAYNE

SHERIDAN HOUSE

This edition first published 2003 by
Sheridan House Inc.
145 Palisade Street,
Dobbs Ferry, NY 10522

Library of Congress Cataloging-In-Publication Data

Payne, John C.
    Understanding boat batteries and battery charging /
John C. Payne.
        p. cm.
    ISBN 1-57409-162-X (alk. paper)
        1. Storage batteries. 2. Boats and boating-Electric equipment
    I. Title.
    VM367.S7 P39 2003
    623.8'503-dc21   2002154369

Printed in the United States of America
ISBN 1-57409-162-X

Contents

1. BATTERIES

What is the purpose of a boat battery?

The battery is the power storage device that is used to start the engine or to power lights and accessories, such as radios, pumps and electronics. The battery also acts as a "buffer" which absorbs the power surges and disturbances that occur during battery charging and discharging.

What type of battery is required?

The type of battery required depends on the service or power discharge requirements of the boat. The service requirements can be grouped under house power, deep cycle or service loads; the other requirements are starting or high current loads. When the load type is decided, the type of battery chemistry can be chosen. This may be the lead-acid flooded cell; the gel cell; the AGM or alkaline type battery.

What is a service, house power or deep-cycle load?

Deep-cycle, service or house power loads are those loads that draw current over long periods of time. Equipment in this category includes the cabin lights, refrigeration, electronic instruments, radios, radar, autopilots, inverters, trolling motors and entertainment systems. The deep-cycle battery is normally used to supply these applications. Calculations are based on the maximum power consumption over the expected longest period between battery recharging.

What is a starting or high current load?

Starting loads require large current levels for relatively short periods of time such as the engine starting motor. Loads in this category also include diesel engine pre-heaters or glow-plugs, the anchor windlass, electric winches, electric thrusters and electric toilets. The starting type battery is normally used for these applications. The battery rating should allow for worst case scenarios, such as very cold temperatures. In cold temperatures the battery efficiency is lowered and engine starting requires greater power due to increased oil viscosity.

When are batteries connected in series?

Batteries are connected in series to increase the voltage. Batteries and cells with ratings of 1.2V, 6V, 8V and 12V can be connected in series to make up banks of 12V, 24V, 32V, 36V or 48V. Six-volt batteries are frequently used in 12- and 24-volt systems as they are easier to install and remove. When connecting batteries in series, the batteries should be of the same make, rating, model and age. If one battery requires replacement, the other should also be replaced. In some installations a series-parallel switch or relay is used to connect batteries for 24 or 48 volts to power thrusters and trolling motors.

Series Connection

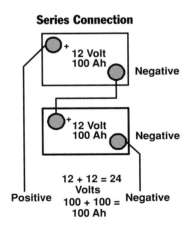

When are batteries connected in parallel?

Batteries are connected in parallel to increase the rating or amp-hour power capacity for the same voltage. Batteries up to around 115Ah are a popular arrangement. They are usually installed in parallel banks of up to three. It is common to have large traction or truck batteries of very large dimensions such as the 8D size installed. As they are physically very large and heavy, servicing can be difficult although manufacturers are trying to solve this problem.

Parallel Connection

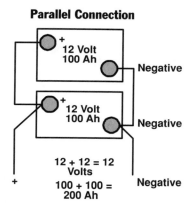

About connecting batteries

When connecting batteries, quality battery terminals must be used. The wing nuts type should not be used as they are hard to tension up without breaking the ears or wings off. Always use a bolt- or clamp-type connector. The clamp type does not require a terminal lug. It is also good protection to install a terminal cover over the connection to prevent accidental contact.

Battery Terminals

Battery Terminal Covers

2. LEAD ACID BATTERIES

How does a flooded or wet cell lead acid battery work?

When two electrodes of different metal are placed in an electrolyte, they form a galvanic cell. An electrochemical process takes place within each cell which generates a voltage. In the typical flooded lead-acid cell the generated voltage is nominally 2.1 volts per cell. In a normal flooded cell lead-acid battery, water loss will occur when it is electrically broken down into oxygen and hydrogen near the end of charging. In normal batteries, the gases disperse to the atmosphere, resulting in electrolyte loss. These are the bubbles seen in the cells during charging.

What are the components of a battery cell?

The typical 12-volt battery consists of 6 cells, which are internally connected in series to make up the battery. The battery cell is made up of several basic components.

- Lead dioxide ($PbO2$), which is the positive plate active material and is brown in color.

- Lead peroxide (Pb), which is the negative plate material and is grey in color.

- Sulfuric acid ($H2SO4$), which is the electrolyte in a diluted solution with water.

- The grids, which are used to hold the lead dioxide and lead peroxide plate material.

- The separators, which are used to hold or space the plates apart.

- The battery casing, which is used to contain each cell.

- The terminals, which are used to connect the cells.

What is Specific Gravity?

The cell electrolyte is the battery acid. This is a dilute solution of sulfuric acid and water. Specific Gravity (SG) is the measurement used to define electrolyte acid concentration. A fully charged cell has an SG typically in the range 1.240 to 1.280, corrected for temperature. This is an approximate volume ratio of acid to water of 1:3. Pure sulfuric acid has an SG of 1.835, and water has a nominal 1.0. The SG of a battery also indicates charge level.

Lead Acid Battery State of Charge Table

S.G.	Voltage	Charge Level
1250	12.75	100%
1235	12.65	90%
1220	12.55	80%
1205	12.45	70%
1190	12.35	60%
1175	12.25	50%
1160	12.10	40%
1145	11.95	30%
1130	11.85	20%
1115	11.75	10%
1100	11.65	0

What happens when a cell discharges?

When an external load such as a light is connected across the positive and negative terminals, the cell will discharge. As there is a potential or voltage difference between the two poles, electrons will flow from the negative pole to the positive pole. A chemical reaction then takes place between the two plate materials and the electrolyte. During the discharge reaction, the plates interact with the electrolyte to form lead sulfate and water. This reaction dilutes the electrolyte, reducing the density. As both the plates become similar in composition, the cell loses the ability to generate a voltage.

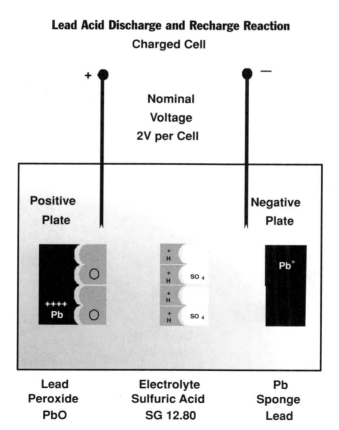

Lead Acid Discharge and Recharge Reaction
Charged Cell

How is a cell recharged?

The charging process reverses the discharge reaction. The water decomposes to release the hydrogen and oxygen. The two plate materials are then restored to the original material. When the plates are fully restored and the electrolyte is returned to the nominal density, the battery is completely recharged.

Lead Acid Charge and Discharge Reaction

Discharged Cell

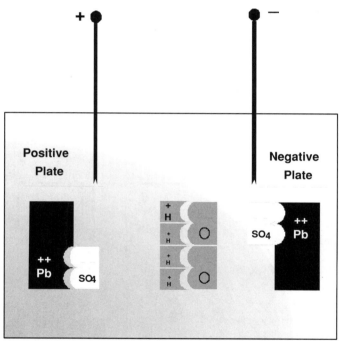

| Lead Sulfate
PbSO4 | Electrolyte
Sulfuric Acid
SG 11.20 | Lead Sulfate
PbSO4 |

How does temperature affect the readings?

For accuracy, all hydrometer readings should be corrected for temperature. Ideally, actual cell temperatures should be used, but ambient battery temperatures are sufficient. Hydrometer floats usually have the reference temperature printed on them and this should be used for calculations. For every 2.7°F (1.5°C) the cell temperature is *above* the reference value, *add* 1 point (0.001) to the hydrometer reading. For every 2.7°F (1.5°C) the cell temperature is *below* the reference value, *subtract* 1 point (0.001) from the hydrometer reading. For example if the reference temperature is 68°F (20°C), at a temperature of 77°F (25°C) add 0.004 to the reading. If the temperature is near freezing at 41°F (5°C), deduct 0.012 from the reading.

How important is battery water?

When topping up the cell electrolyte, always use distilled or de-ionized water. Rainwater is acceptable, but under no circumstances use water out of the galley tap or faucet, as this water generally has an excessive mineral content or other impurities that pollute and damage the cells. Impurities introduced into the cell will remain, and concentrations will accumulate at each top up and reduce service life.

What is plate sulfation?

Plate sulfation is the most common cause of battery failure. When a battery is discharged the chemical reaction converts both plates to lead sulfate. If recharging is not carried out quickly, the lead sulfate starts to harden and crystallize. This is characterized by white crystals on the brown plates and is almost non-reversible. The immediate effect of sulfation is partial and permanent loss of capacity as the active plate materials are reduced. Electrolyte density also partially decreases, as the chemical reaction during charging cannot be fully reversed. This sulfated material introduces higher resistances within the cell and inhibits charging. As the level of sulfated material increases, the cell loses the capability to retain a charge and the battery fails.

How efficient is a battery?

Battery efficiency is affected by temperature. At 32°F (0°C), the efficiency will fall by 60%. Batteries in warm tropical climates are more efficient, but may have reduced life spans, and batteries commissioned in tropical areas often have lower acid densities. Batteries in cold climates have increased operating lives, but are less efficient.

What is self discharge?

During charging, small particles of the antimony used in the plates and other impurities dissolve out of the positive plates and deposit on to the negative ones. Other impurities from the topping up water also deposit on the plates. A localized chemical reaction then takes place, slowly discharging the cell. Self-discharge rates are affected by temperature. At 32°F (0°C), the self-discharge rates are small. At 86°F (30°C), the self-discharge rates are very high and the specific gravity can decrease by as much as 0.002 per day, typically up to 4% per month.

What do the group numbers mean?

These are standard sizes used in the United States. The following are for 12-volt batteries that are typically used in boats and will vary according to battery type. Large 8D batteries are now easier to handle with Rolls having cell based units.

Group	Capacity
24	65-75 Ah
27	80-90 Ah
31	105 Ah
904D	160 Ah
908D	225 Ah
4D	160-200 Ah
8D	220-245 Ah

What is a dual purpose battery?

These batteries are a combination of deep cycle and starting battery. They have both the high cranking capacity with good cycling ability. They are used in many smaller boats where smaller house loads do not require a separate deep cycle battery and a start battery.

When should a battery be recharged?

A deep cycle battery should be recharged when the charge level falls to a maximum of 50%. A starting battery should always be recharged immediately after each use and maximum allowable level of 10% of maximum charge level. If a start battery is deep cycled, the life of the battery will be severely reduced.

Rolls Red Battery
Courtesy Rolls

Mastervolt Battery
Courtesy Mastervolt

3. AGM BATTERIES

What are Absorbed Glass Mat (AGM) or Valve Regulated Lead Acid (VRLA) batteries?

These are new generation battery types. The electrolyte is held within a very fine microporous, boron-silicate glass matting that is placed between the plates. This absorbs and immobilizes the acid while still allowing plate interaction. They are also called starved electrolyte batteries, as the mat is only 95% soaked in electrolyte.

AGM Battery Principles

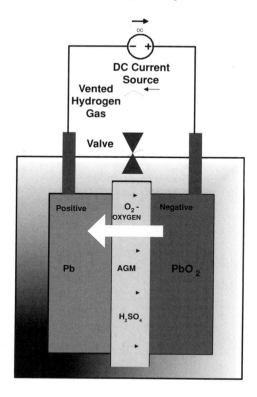

How does an AGM battery work?

These batteries use a principle called Recombinant Gas Absorption. The plates and separators are held under pressure. During charging, the evolved oxygen is only able to move through the separator pores from positive to negative, reacting with the lead plate. The negative plate charge is then effectively maintained below 90%, which stops hydrogen generation.

How are AGM batteries charged?

Typical charge voltages are in the range 14.4 to 14.6 volts at 68°F (20°C). The batteries have a very low internal resistance, which results in minimal heating effects during heavy charge and discharge. They can be bulk charged at very high currents, typically by a factor of five over flooded cells, and a factor of 10 over gel batteries. They also allow 30% deeper discharges and recharge 20% faster than gel batteries, and have good recovery from full discharge conditions. Self-discharge rates are only 1%-3%. Typical charge voltage levels are: 100% is 12.8-12.9 volts, 75% is 12.6 volts, 50% is 12.3 volts, 25% is 12 volts and 11.8 volts is flat. At high temperatures both AGM and gel cells are unable to dissipate the heat generated by oxygen and hydrogen recombination and this can create thermal runaway. This will lead to gassing and the drying out of cells. A premature loss of capacity can occur when the positive plate and grids degrade due to higher operating temperatures caused by the recombination process and higher charge currents.

4. GEL BATTERIES

What is a gel battery?

The gel cell has a solidified gel as an electrolyte, which is locked into each group of plates. During charging the gel liquefies due to its thixotropic properties, and solidification after charging can exceed an hour as thixotropic gels have a reduced viscosity under stress. The newer battery types use phosphoric acid in the gel to retard the sulfation hardening rates.

How is a gel battery constructed?

The lead plates in a gel cell are reinforced with calcium, rather than the antimony used in flooded cells. This reduces self-discharge rates, and they are relatively thin. This also helps gel diffusion and improves the charge acceptance rate as diffusion problems are reduced. The separator provides electrical and mechanical isolation of the plates. Each cell has a safety valve to relieve the excess pressure if the set internal pressure is exceeded. The valve will then reclose tightly to prevent oxygen from entering the cell.

Gel Cell Battery Principles

Vented Gas

Valve

Positive	O_2	Negative
Pb	O_2 O_2 O_2	PbO_2

Gel electrolyte — Separator

How are gel cell batteries charged?

The gel battery has a higher charge acceptance rate than the flooded cell battery. This allows a more rapid charge rate and the typical rate is 50% of Ah capacity. A gel cell battery cannot tolerate having any equalizing charge applied; this over charge condition will seriously damage them. During charging the current causes the decomposition of the water and the generation of oxygen at the positive plate. The oxygen diffuses through the unfilled glass mat separator pores to the negative plate, and reacts chemically to form lead oxide, lead sulfate and water. The charge current then reduces and does not evolve hydrogen gas. If recombination of hydrogen is incomplete during overcharge conditions, the gases may vent to the battery locker. While these batteries will accept some 30-40% greater current than an equivalent lead acid battery they are restricted in the voltage levels allowed, so you cannot use any fast charging system. Typical open circuit voltages are: 100% charge is 12.85-12.95 volts, 75% is 12.65 volts, 50% is 12.35 volts, 25% is 12 volts and 11.8 volts is flat. The normal optimum voltage tolerance on Dryfit units is 14.4 volts. There are some minimal heating effects during charging caused by the recombination reaction. Continuous over- or undercharging of gel cells is the most common cause of premature failure. In many cases this is due to use of imprecise automotive type chargers.

5. GENERAL INFORMATION ON BATTERY TYPES

What is an alkaline cell battery?

Alkaline cells are also commonly known as Nickel Cadmium (NiCad) and Nickel Iron (NiFe) batteries. The principal factors are cost (typically 500% more), greater weight and physical size, and they are normally only be found in larger boats. The batteries use an alkaline electrolyte and not an acid. Unlike lead-acid cells, plates undergo changes in their oxidation state, and do not alter physically. As the active materials do not dissolve in the electrolyte, the plate life is very long. The positive plate is made of Nickel-Hydroxide, and the negative plate of Cadmium Hydroxide. The electrolyte is a potassium hydroxide solution with a specific gravity of 1.3. Unlike lead-acid cells, the density does not significantly alter during charge and discharge and hydrometer readings cannot be used to determine the state of charge. Electrolyte loss is relatively low in operation. Lead-acid and NiCad batteries should never be located in the same compartment as the cells will become contaminated by acid fumes causing permanent damage.

What are the advantages and disadvantages of battery types?

Flooded cell batteries cost less, they are lighter in weight, and are more resilient to overcharge conditions. The disadvantages are that maintenance is required, acid can be spilled, gases are generated, and they have relatively high self-discharge rates.

AGM batteries don't require maintenance, and don't spill acid or generate gas in normal operation. They also have low self-discharge rates, have good deep-cycle ability and maintain a relatively constant voltage during discharge. The disadvantages are that they are expensive, and if overcharged they are permanently damaged, and they are heavy.

Gel batteries don't require maintenance, don't spill acid or generate gas in normal operation. They also have low self-discharge rates, and some have good deep-cycle ability. The disadvantages are that they are expensive, and if overcharged they are permanently damaged.

Can I mix the battery types?

Battery types must not be mixed. Each battery type has a different charging voltage and discharge characteristic, so the battery types must be the same. Also never mix old batteries with new batteries.

What are the life expectancies of each battery type?

Numbers are usually based on the manufacturer's data. A quality deep-cycle lead acid traction battery can have a life exceeding 2500 cycles of charge and discharge to 50%. A gel cell has a life of approximately 500-800 cycles depending on the make. An AGM battery has a cycle life of 350 to 2200 depending on the type.

What happens when a battery is completely flattened?

Deep-cycle batteries are permanently damaged within days by sulfation of plates. Gel batteries can survive up to a month, while AGM can survive also about a month. All can be recharged however damage occurs and cycle life is reduced.

6. BATTERY RATINGS AND SELECTION

What is the Amp-hour rating of a battery?

The Amp-hour rating (Ah) of a battery refers to the available current a battery can deliver over a nominal period until a specified final voltage is reached, or amps per hour. Amp-hour rates are normally specified at the 10- or 20-hour rate. This rating is normally only applied to deep-cycle batteries. For example, a battery is rated at 105 Ah at 10-hour rate, final voltage 1.7 volts per cell. This means that the battery is capable of delivering 10.5 amps for 10 hours, when a cell voltage of 1.7 volts will be reached. At this point the battery volts will be at 10.2 volts.

What is the Reserve Capacity rating of a battery?

The Reserve Capacity rating specifies the number of minutes a battery can supply a nominal current at a nominal temperature without the voltage dropping below a certain level. It indicates the power available when an alternator fails and the power available to operate ignition and auxiliaries. Typically, the rating is specified for a 30-minute period at 77°F (25°C) with a final voltage of 10.2 volts.

What is the Cold Cranking Amps (CCA) rating?

The Cold Cranking Amp rating specifies the current available at 0°F (-18°C) for a period of 30 seconds, while being able to maintain a cell voltage exceeding 1.2 volts per cell. This rating is only applicable for engine starting. The higher the battery rating, the more power available, especially in cold weather conditions.

What is the Marine Cranking Amps (MCA) rating?

The Marine Cranking Amp rating is a relatively new rating which defines the current available at 32°F (0°C) for a period of 30 seconds, while being able to maintain a cell voltage exceeding 1.2 volts per cell. This rating is only applicable for engine starting. If you are in cold climate area such as the UK, Europe, United States or Canada, the CCA rating is more relevant.

Why are plate numbers quoted?

Many battery data sheets specify the number of plates. This is defined as the number of positive and negative plates within a cell. The more plates installed, the greater the plate material surface area. This increases the current during high current rate discharges and the cranking capacity and cold weather performance are improved.

What is a marine battery?

This is a sales or marketing description for certain constructional features. The plates may be thicker than normal or there may be more of them. The internal plate supports may be able to absorb vibration. The battery cases may be manufactured with a resilient rubber compound. The cell filling caps may be of an anti-spill design. In general, batteries are of a similar design with very little to distinguish the automotive types, except the label.

What is a deep-cycle battery?

Service loads require a battery that can withstand cycles of long continuous discharge, and repeated recharging. This deep cycling requires the use of the suitably named deep-cycle battery. These are available in all battery chemistry types.

How is the deep-cycle battery constructed?

The lead acid deep-cycle battery is typified by the use of thick, high-density flat-pasted plates, or a combination of flat and tubular plates. The plate materials may also contain small proportions of antimony to help stiffen them. Porous, insulating separators are used between the plates and glass matting is used to assist in retaining active material on the plates that may break away as they expand and contract during charge and recharge cycles. If material accumulates at the cell base, a cell short circuit may occur, although this is less common in modern batteries. If material is lost, the plates will have reduced capacity or insufficient active material to sustain the chemical reaction with resultant cell failure. Much has been done to develop stronger and more efficient plates. Rolls have their Rezistox positive plates. The grid design has fewer heavier sections to hold the high density active material. This is due to the dynamic forces that normally cause expansion and contraction with subsequent warping and cracking. Separator design has also evolved; Rolls use double insulated thick glass woven ones that totally encase the positive plate along with a microporous polyethylene envelope. This retains any material shed from the plates than cause cell short circuits.

How many deep cycles are available?

The number of available deep cycles varies between the different battery makes and models. Typically it is within the range of 800-1500 cycles of discharge to 50% of nominal capacity and then *complete* recharging. Battery life is a function of the number of cycles and the depth of cycling. Batteries discharged to only 70% of capacity will last much longer than batteries that are discharged to 40% of capacity. In practice, plan your system so that discharge is limited to 50% of battery capacity. The typical life of batteries where batteries are properly recharged and cycle capabilities maximized can be up to 5-10 years.

Deep Cycles
Courtesy Surrette

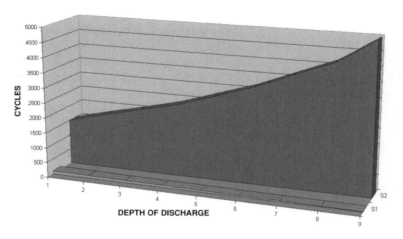

How is a deep cycle battery selected?

Many onboard power problems result from improperly selected batteries. Battery bank capacities can be seriously underrated, which causes power shortages. If batteries are overrated the charging system cannot properly recharge them, and sulfation of the plates can occur with premature failure. All the electrical equipment on board is listed along with the power consumption ratings.

What happens if the battery is discharged faster than the nominal rate?

The faster a battery is discharged over the nominal rating, which is either the 10- or 20-hour rate, the less the real amp-hour capacity the battery has. This effect is defined by Peukerts Equation, which has a logarithmic characteristic. This equation is based on the high and low discharge rates and discharge times for each to derive the Peukert coefficient 'n'. Average values are around 1.10 to 1.20. If we discharge a 250 amp-hour battery bank, which has nominal battery discharge rates for each identical battery of 12 amps per hour at a rate of 16 amps, we will actually have approximately 10-15% less capacity. Battery discharge meters such as the E-Meter incorporate this coefficient into the monitoring and calculation process.

What happens when the battery is slowly discharged?

The slower the battery is discharged over the nominal rate, the greater the real battery capacity. If we discharge our 240 amp-hour battery bank at 6 amps per hour we will actually have approximately 10-15% more capacity. The disadvantage here is that slowly discharged batteries are harder to charge if deep cycled below 50%.

Can all the battery capacity be used?

For a deep-cycle battery the discharge capacity should be kept at 50% of nominal battery capacity to maximize life. We can assume that a battery capacity of 240 amp-hours is the basic minimum level. This should be a minimum requirement, but certain frightening realities must now be introduced into the equation. The figures below typify a common system, with alternator charging and a standard regulator. Maximum charge deficiency is based on the assumption that boat batteries are rarely above 70% charge and cannot be fully recharged with normal regulators. There is reduced capacity due to sulfation, which is typically a minimum of 10% of capacity. The key to maintaining the optimum power levels and avoiding this common and surprising set of numbers is the charging system.

Nominal Capacity	240 Ah
Maximum Cycling Level (50%)	Deduct 120 Ah
Max. Charge Deficiency (30%)	Deduct 72 Ah
Lost Capacity (10%)	Deduct 24 Ah
Available Battery Capacity	**24 Ah**

What is a discharge requirement?

The nominal required battery capacity of 240 Ah has been calculated as that required to supply 10 amps per hour over 12 hours to 50% of battery capacity. In most cases, the discharge requirements are worst for the night period, and this is the 12-hour period that should be used in calculations. What is required is a battery bank with similar discharge rates as the current consumption rate. This will maximize the capacity of the battery bank with respect to the effect defined in Peukerts coefficient.

What is battery load matching?

The main aim is to match the discharge characteristics of the battery bank to that of our calculated load of 10 amps per hour over 12 hours. Calculations assume that the charging system is able to recharge batteries to virtually 100% of nominal capacity. The factors affecting matching are the discharge and battery capacity requirements.

About battery capacity requirements

As the consumption rate is based on a 12-hour period, a battery bank that is similarly rated at the 10-hour rate is required. In practice you will not match the precise required capacity, therefore you should go to the next battery size up. This is important also as the battery will be discharged longer and faster over 12 hours, so a margin is required. If you choose a battery that has 240 amp-hours at the 20-hour rate, in effect you will actually be installing a battery that in the calculated service has 10-15% less capacity than that stated on the label, which will then be approximately 215 Ah, so you are below capacity. This is not the fault of the supplier, but simply a failure to correctly calculate and specify the right battery to meet system requirements.

The ratings can usually be found on the equipment nameplates or in the equipment manuals. The ratings are given in amps (A) or watts (W). It is recommended that where watts are used, this rating is converted to current in amps. To do this, divide the power rating in watts by the operating voltage, 12 or 24 volts.

Watts (W)	Amps (A)
6 watts	0.5 amp
12 watts	1.0 amp
18 watts	1.5 amps
24 watts	2.0 amps
36 watts	3.0 amps
48 watts	4.0 amps
60 watts	5.0 amps
72 watts	6.0 amps

What time periods should be used in the power consumption calculation?

For sailing or power boats, calculate the power consumption for a 12-hour period while in port or at anchor. The calculation should assume that the engine will not be operated, and that no generator or shore power with battery charger will be operational. When a boat is motoring, all the power is being supplied from the engine alternators, and when the batteries are fully charged, the alternator effectively supplies all power.

What is a Load Calculation Table?

A Load Calculation Table is used to list and carry out calculations. To calculate the total boat electrical system loading, multiply the total current values by the number of operating hours to get the amp-hour rating. If equipment uses a current of 1 amp over a 24-hour period, then it consumes 24 amp-hours (Ah). The table shows many typical power consumptions. There is space for insertion and calculation of your own boat electrical data.

How to perform a load calculation

All equipment on the boat must be listed along with the power consumption ratings. Ratings are found on the equipment nameplates and in equipment manuals. Insert your own values into the current column, typical values are in brackets. Calculate the power used over 12 hours. To convert power in watts to the current in amps, divide the power value by the system voltage. Add up all the current figures relevant to your vessel and multiply by hours (12) to determine the average amp-hour consumption rate. Most equipment will be on when sailing, anchored or moored. Equipment used when motoring or when shore power is connected is not relevant. Depending on the frequency between charging periods select the column that suits your boat activity. The most typical scenario is where the boat is moored or at anchor and the lights, entertainment systems, some radio or navigation equipment and an electric refrigerator are used.

Eg. Total consumption is 120 Ah over 12 hours = 10 amps/hour

Load Table 1 - DC Load Calculation

Equipment	Current (A)	Consumption (Ah)
Radar -Transmit	(4.5A)	
Radar - Standby	(0.5A)	
VHF - Receive	(0.5A)	
Weatherfax	(0.5A)	
GPS	(0.5A)	
Navtex	(0.5A)	
Fishfinder	(1.5A)	
Instruments	(0.5A)	
Stereo	(1.0A)	
Inverter Stby	(0.5A)	
Anchor Light	(1.0A)	
Refrigeration	(4.0A)	
Interior Lights	(4.0A)	
Computer	(2.5A)	
Television	(2.0A)	
Video	(1.5A)	
Trolling Motor	(15.0 A)	

How should intermittent loads be calculated?

Some loads come on and off periodically such as water pumps, using battery power for short periods. It is difficult to quantify actual real current demands with intermittent loads. I use a baseline of 6 minutes per hour, which is .1 of an hour.

Load Table 2 - Intermittent DC Load Calculation

Equipment	Current (A)	Consumption (Ah)
Bilge Pump	(3.5A)	
Water Pump	(3.5A)	
Shower Pump	(2.5A)	
Toilet	(15.0A)	
Cabin Fans	(1.0A)	
VHF Transmit	(4.0A)	
Spot Light	(3.0A)	
Inverter	(40.0A)	
Cabin Lights	(4.0A)	
Load Table 1		
Load Table 2		
TOTAL LOAD		

How to select start batteries?

The starting battery must be capable of delivering the engine starter motor with sufficient current to crank over and start the engine. This starting load can be increased by engine compression, oil viscosity and engine driven loads. Loads such as a thruster or anchor windlass under full load also require similar large amounts of current. The starting battery is normally selected on the basis of the engine manufacturer's recommendations and the table shows recommended battery ratings for various diesel ratings as well as typical starter motor currents. It is good practice to allow for a multi-start capability as a safety margin. Allowances should be made for the decreased efficiency of engines in cold climates as a greater engine starting current is required, and therefore larger battery ratings.

Battery Ratings Table

Engine Rating	Current Load	Battery CCA
10 Hp 7.5 kW	59 amps	375 CCA
15 Hp 11 kW	67 amps	420 CCA
20 Hp 15 kW	67 amps	420 CCA
30 Hp 22 kW	75 amps	450 CCA
40 Hp 30 kW	85 amps	500 CCA
50 Hp 37 kW	115 amps	500 CCA
100 Hp	115/60 amps	500 CCA
150 Hp	150/75 amps	600 CCA
200 Hp	120 amps	800 CCA

How a starting battery is constructed

The starting battery contains relatively thin, closely spaced porous plates, which gives maximum exposure of active plate material to the electrolyte with minimal internal resistance. This enables maximum chemical reaction rates, and maximum current availability.

Can start batteries be deep cycled?

Starting batteries cannot withstand cycling, and if deep cycled or flattened have an extremely short service life. Ideally they should be maintained within 95% of full charge. Cycle life can be as low as 25-50 cycles. Start batteries if not fully recharged will suffer from sulfation. If improperly used for deep cycle applications and under charged they will sulfate.

Will start batteries self discharge?

Starting batteries have very low self-discharge rates and this is generally not a problem in normal engine installations.

How efficient are start batteries?

Cold temperatures dramatically affect battery performance. Engine lubricating oil viscosities are also affected by low temperatures, and further increase the starting loads on the battery.

Battery Power Table

Temperature	Battery Level	Power Required
80°F (+27°C)	100%	100%
32°F (0°C)	65%	155%
0°F (-18°C)	40%	210%

How should start batteries be recharged?

The discharged current must be restored quickly to avoid damage, and charge voltages should compensate for temperature. Normally after a high current discharge of relatively short duration, there is no appreciable decrease in electrolyte density. The battery is quickly recharged, as the counter voltage phenomenon does not have time to build up and has a negligible effect on the charging.

7. SAFETY, INSTALLATION AND MAINTENANCE

How hazardous are batteries?

The lead-acid battery is potentially hazardous and proper safe handling procedures should be followed. Battery cells contain an explosive mixture of hydrogen and oxygen gas at all times. An explosion risk exists at all times if naked flames, sparks or cigarettes are introduced into the immediate vicinity. Always use insulated tools. Cover the terminals with an insulating material to prevent accidental short circuit. Watchbands, bracelets and neck chains can accidentally cause a short circuit. Sulfuric acid is highly corrosive and must be handled with extreme caution. Always lift the battery with carriers if fitted. If no carriers are fitted, lift using opposite corners to prevent case distortion and electrolyte spills. If a fast charging device is installed, ensure that the ventilation remains sufficient to remove any generated gasses, and prevent them from accumulating.

Where should batteries be installed?

Batteries should be installed within a separate space or compartment that is located above the maximum bilge water level, and protected from mechanical damage. The batteries should be installed in a lined box protected from temperature extremes, although plastic containers are commonly used. The preferred temperature range is 50°-80°F (10°C-27°C). Always allow sufficient vertical clearance to install and remove batteries, as well as access and allow hydrometer testing. The batteries should never be installed close to any source of ignition, such as fuel tanks and system parts such as fuel filters, separators and valves. Any leak or accumulation of fuel represents a serious hazard and so any source of ignition should be removed.

What about battery boxes?

The battery box should be made of plastic, fiberglass or lead lined to prevent any acid spills coming into contact with wood or water. Boxes should be at least the full height of the battery so that any spills will be contained at all times. PVC battery boxes are acceptable alternatives. Battery boxes must also be fastened down and tie-down straps are available. Physically secure batteries with either straps or a removable restraining rod across the top to prevent any movement. Insert rubber spacers around the batteries to stop any minor movements and vibrations. Battery box lids should be in place at all times and secured. PVC or other connection covers should be installed where accidental contact by tools or other items can cause a short circuit across the terminals. Terminals should be coated to limit the corrosive effects of acid.

Battery boxes
Courtesy Blue Sea Systems

How should battery negatives be interconnected?

When two or more batteries are connected in parallel, all the battery negatives are also connected together. When a house battery bank and a starting battery are charged from the same source, the negatives are also connected. The polarizing ground is then connected to one of the battery negatives. In dual battery systems the cables connecting each battery negative or positive should be rated the same.

How should battery positives be connected?

The positive cables must be connected in such a way that the start battery cannot not be accidentally discharged. Where a solenoid system is used to parallel the batteries for charging, it must always open when the charging stops. Where high current equipment can cause system disturbances such as large load surges and voltage droops, consideration should be given to installing a separate battery bank with the required characteristics to power the equipment.

Parallel Connection Unit
Courtesy Blue Sea Systems

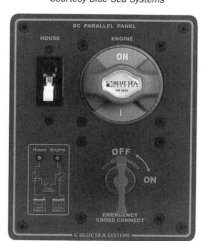

What battery maintenance is required?

Check the density of the electrolyte but do not test immediately after charging or discharging. Wait at least half an hour until the cells stabilize, as it takes some time for the pockets of varying electrolyte densities to equalize. Never test immediately after topping up the electrolyte. Wait until after a charging period, as it similarly takes times for the water to mix evenly. Check the electrolyte level in each cell. Always top up electrolyte levels with deionized or distilled water to the correct levels. Ensure that the terminal posts are clean and tight. Coat the terminals with petroleum jelly. Clean the battery surfaces with a clean, damp cloth. Moisture, salt and other surface contamination can cause surface leakage between the positive and negative terminals across the battery casing top. This can slowly discharge the battery and is a common cause of flat batteries, and mysterious but untraceable system leaks.

What is a load test?

A fully charged battery has a load connected along with a digital voltmeter. Typically the load is rated at about half the rated CCA value and is often around 300 amps. This is connected for 15 seconds. The voltage at the end of this period is an indicator of condition and the capability to sustain the electrical chemical reaction. A reading of over 9.6 volts indicates that the battery is still serviceable; a rating under 9.6 volts indicates the battery is failing.

8. CHARGING, ALTERNATORS AND REGULATORS

About charging efficiency

Manufacturers specify nominal capacities of batteries, and the total capacity of the bank must be taken into consideration. Older batteries have reduced capacities due to normal in-service aging, and plate sulfation, which increases internal resistance and also inhibits the charging process. The electrolyte is temperature dependent, and the temperature is a factor in setting maximum charging voltages. The state of charge at charging commencement can be checked using the open circuit voltage test and electrolyte density. The level of charge will affect the charging rate.

What is the correct charging voltage?

Charging voltage is defined as the battery voltage plus the cell voltage drops. Cell volt drops are due to internal resistance, plate sulfation, electrolyte impurities and gas bubble formation that occurs on the plates during charging. These resistances oppose the charging and must be exceeded to effectively recharge the battery. Resistance to charging increases as a fully charged state is reached and decreases with discharge. A battery is self-regulating in terms of the current it can accept under charge. Over-current charging at excessive voltages simply generates heat and damages the plates.

What does bulk charge mean?

The bulk charge phase is the initial charging period where charging takes place until the gassing point is reached. This is typically in the range 14.4 to 14.6 volts, corrected for temperature. The bulk charge rate can be anywhere between 25% and 40% of rated amp-hour capacity at the 20-hour rate as long as temperature rises are limited.

What does absorption charge mean?

After attaining the gassing voltage, the charge level should be maintained at 14.4 volts until the charge current falls to 5% of battery capacity. This level is normally 85% of capacity. In a typical 300 amp-hour bank, this is 15 amps.

What is a float charge?

The battery charge rate should be reduced to a float voltage of approximately 13.2 to 13.8 volts to maintain the battery at full charge.

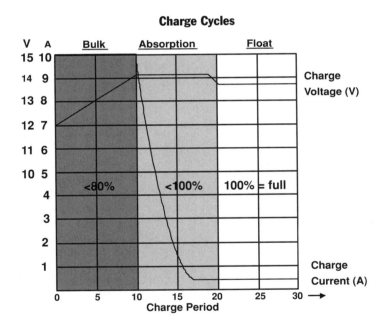

Charge Cycles

What is an equalization charge?

An equalization charge consists of applying a higher voltage level at a current rate of 5% of battery capacity in order to "re-activate" the plates. This will not completely reverse the effects of sulfation, and there may be an improvement following the process, but it will not reverse long-term permanent damage. Equalization at regular intervals can increase battery longevity by ensuring complete chemical conversion of plates, but care must be taken. Equalization charges are typically set at 16.5 volts for up to 3 hours; it is essential that all circuits be off at the switch panel so that higher voltages cannot damage equipment power supplies.

What is the right charging capacity?

From the power analysis table we have calculated the maximum current consumption. Added to this is a 20% margin for battery losses giving a final charging value. A battery requires the replacement of 120% of the discharged current to restore it to full charge. This value is required to overcome losses within the battery due to battery internal resistances during charging. A popular benchmark is that alternator rating should be approximately 30 to 40% of the battery capacity. As a battery is effectively self-limiting in terms of charge acceptance levels, it is not possible to simply apply the discharged value and expect the battery to recharge. The battery during charging is reversing the chemical reaction of discharge, and this can only occur at a finite rate. The alternator therefore must be selected if possible to recharge at the battery optimum charge rate as specified. Charging has a tapered characteristic, which is why start and finishing rates are specified. The required charging current is the sum of the charge rate plus anticipated loads during charging.

How does an alternator work?

The alternator is driven by a rubber belt from the engine. When the alternator reaches a certain cut-in speed, electrical power will be produced. An excitation current is used to generate a magnetic field in the rotor so that required alternator voltage can be induced into the stator windings. Excitation current goes through the exciter diodes; through the carbon brushes to the collector sliprings and to the excitation winding. This goes to the field (DF) terminal of voltage regulator and then to negative terminal (D-) of voltage regulator, and back to the stator winding through the power diodes. This is then rectified to produce a DC output for charging via the full wave bridge rectifier.

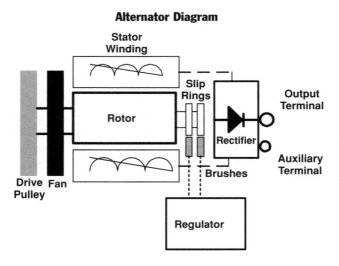

Alternator Diagram

Alternator
Courtesy Volvo

Alternator
Courtesy Mastervolt

What is the alternator stator?

The stator is the fixed winding that creates an electromagnetic field. It consists of a three-phase winding that is connected in a "star" or a "delta" arrangement. The windings are formed onto a solid laminated core. These supply three phases of alternating current (AC) to the rectifier.

What is the alternator rotor?

The rotor is the rotating part of the alternator. The shaft has the pole or claw shaped magnet poles attached, the excitation winding, the cooling fan at one end, the bearings, and the collector sliprings.

What is the alternator rectifier?

The rectifier consists of a network of six diodes, which are connected across the positive and negative plates. Two power diodes are connected in each phase, one diode is connected to the positive side, and one to the negative side.

These plates also function as heat sinks to dissipate the heat from power generation. The positive half-waves pass through the positive side diodes and the negative half-waves pass through the negative diodes. This rectifies the three generated AC phase voltages into the DC output for charging. Two diodes are used on each winding to provide full-wave rectification. The rectifier diodes prevent the battery discharging through the 3-phase winding as the diodes are polarized in the reverse direction.

What are the alternator exciter diodes?

The exciter (D+) or pre-excitation diodes consist of three low power diodes which independently rectify each AC phase and provide a single DC output for the warning light or auxiliary control functions. They are required as the residual magnetism in the iron core is insufficient at low speeds and starting to initiate the self-excitation required to build up the magnetic field. This only occurs when the alternator voltage is higher than the voltage drop across the two diodes. Current flows through the charge indicator lamp, through excitation winding, in the stator, through voltage regulator to ground. This battery current pre-excites the alternator. The warning lamp functions as a resistor and provides pre-excitation current, which generates a field in the rotor. In this respect the power or watts rating of the lamp is important and 2-5 watts is typical. In many cases an alternator will not operate if the lamp has failed and this is because the residual voltage or magnetism has dissipated. Undersized lamps are often characterized by the need to rev the engine to get the alternator to kick in.

What is the alternator brush gear?

The brushes are normally made of copper graphite. The brushes are spring-loaded to maintain correct slipring contact pressure and are soldered to the terminals.

What is the function of the voltage regulator?

The voltage regulator is usually combined with the brush gear or mounted adjacent to it. The field control output of the alternator is connected to one of the brush holders, which then supplies the rotor winding through the slipring. Regulator sensing is normally connected to the D+ output circuit. The regulator maintains a constant voltage output over the entire operating range of the alternator. Earlier electro-magnetic contact type regulators are relatively uncommon now with most being electronic types with no moving parts. The electronic regulator allows precise control with short field switching periods.

How are alternators rated?

Alternators are rated at the current output rating for a given temperature. These may vary between 105°F-122°F (40°-50°C). The ratings for vehicles tend to be lower as they get a large airflow over them; however small engine compartments routinely get to around 105°F-122°F (45°-50°C). The hotter the engine compartment gets above rated temperature the lower the output rating gets.

What is the correct output cable size for an alternator?

The positive charging output cable from the alternator to the battery, or the diode, switch or relay should be rated at the maximum rated alternator current and for a maximum voltage drop of 5%. It should also allow for high ambient temperatures. Many installations are underrated and it is best to oversize cables by around 25%.

What is the correct negative cable size for an alternator?

Most alternator installations use the engine block and the main negative from the battery as the return to the battery. To maintain system separation and minimize voltage drops in the charging circuit, which normally includes the engine block, a separate negative cable should be installed for each installed alternator. The negative cable should be the same size as the positive cable, and be connected at the alternator casing or the negative terminal if installed.

What about switches and fuses?

Switches and fuses should not be installed in any alternator output circuit. Alternator failures caused by unintentional operation of changeover switches are very common. When a switch is opened under load the spike normally destroys the alternator diodes. When terminating alternator cables, make sure that crimp connections have the right current rating and the rings have the correct size to match the termination bolts on the alternator and battery. Do not solder connections, as soldered connections frequently fail or are high resistance points in the circuit. After charging physically touch the alternator output terminal; if it is very hot, the connection is probably undersized and therefore overheating, causing charging system power losses.

What is a marine alternator?

Typically marine alternators have over-rated diodes in the rectifier, and larger cooling fans to increase air flow. The windings are protected to a higher standard by epoxy impregnation and the output characteristics are generally similar to automotive types. Manufactured marine units also have a corrosion resistant paint finish and are designed for higher ambient operating temperatures. The bearings should be of the totally enclosed type, and some may use high temperature grease.

What are ignition proof alternators?

These alternators are completely enclosed, with enclosed brushes and sliprings, and ignition protected with a UL listing. This prevents accidental ignition of hazardous vapors in gasoline-powered engines. Never install a normal alternator on a gasoline-ignition engine.

What alternator output rating is required?

The typical rule quoted is 30-40% of the installed battery capacity. If you have 300 Ah, a 100-amp output rating is required. The capacity can also be based on calculating the maximum probable recharge rate that may occur.

How to install alternators correctly

Optimum service life and reliability can only be achieved by correctly installing the alternator. The following factors must be considered during installation.

- **Alignment.** The alternator drive pulley and the engine drive pulley must be correctly aligned. Any misalignment of the pulleys can cause twisting and friction on the drive belts and additional side loading on bearings. Both can cause early failure.

- **Drive Pulleys.** The drive pulleys between the alternator and the engine must be of the same cross-section. Any differences will cause the belt to overheat and early failure. Solid pulleys of the correct ratio should replace the thin split automotive type pulleys that are on some alternators.

- **Drive Belt Tension.** Belts must be correctly tensioned. The maximum deflection must not exceed 12 mm. When a new belt is fitted, the deflection should be re-adjusted after 1 hour of operation and then again after 10 hours as belts will stretch in during this period.

What happens if drive belts are under-tensioned?

Under-tensioning causes the belt to overheat and stretch, and the slipping causes undercharging. The excess heat generated also heats up the pulleys and the high heat level conducts along the rotor shaft to the bearing, melting bearing lubricating grease and increasing the risk of premature bearing failure. You will often hear a screeching sound when speed is applied to the engine.

What happens if drive belts are over-tensioned?

Over-tensioning causes excessive alternator bearing side loads which leads to early bearing failure. Signs of this condition are sooty looking deposits around the belt area, and wear on the edges of the belt.

Which drive belt types are best?

Drive belts must be of the correct cross section to match the pulleys. Notched or castellated belts are a good choice in the hot engine area as they dissipate heat well. If multiple belts are used, always renew all belts together to avoid varying tensions between them. In any alternator rated over 80 amps, a dual belt system should be used. A single belt will not be able to cope with the mechanical loads applied at higher outputs.

How much ventilation is required?

Vehicle alternators have a constant high airflow over them, and boat engine compartments are small and confined with no airflow. An alternator is effectively de-rated in high temperatures similar to electrical cable. Many alternator failures occur after a new fast charging systems is installed as the alternators run at close to maximum output for a period in high ambient temperatures.

How important are alternator mountings?

Alternator brackets are a constant source of failure. When tensioning the alternator, always fix both the adjustment and the pivot bolts. Failure to tighten the pivot bolt is common and this causes the alternator to twist and vibrate. Vibration fatigues the bracket or mounting and causes metal fractures. The mountings are part of the negative return path, and if loose can cause undercharging and radio interference from sparking at the loose connection.

About alternator drive pulley selection

Ideally maximum alternator output is required at a minimum possible engine speed. This is a few hundred revs/min above idle speed. Alternators have three speed levels that must be considered and the aim is to get full output at the lower speeds.

Alternator Test Specifications

Rating	Speed	Output
14V 35A	1300 rpm	10A
	2000 rpm	23A
	6000 rpm	35A
14V 55A	1200 rpm	16A
	2000 rpm	36A
	6000 rpm	55A
12V 130A	1000 rpm	30A
	2000 rpm	115A
	3000 rpm	130A

What are alternator output characteristics?

There is a direct relationship between the alternator output current, efficiency, torque, HP (kW) and the alternator speed. A graph is used to show these characteristics and the ideal speed can be selected from these characteristics. The speed at which an alternator starts generating and where full output is available is easy to determine.

Alternator Output Curve

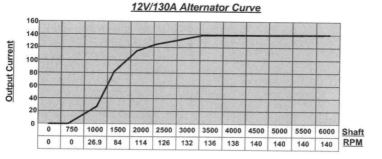

12V/130A Alternator Curve

What are alternator speed ratings?

The speed ratings are referred to as the cut-in speed which is the speed at which a voltage will be generated. The full output operating speed is the speed where full rated output can be achieved. The maximum output speed is the maximum speed allowed for the alternator, otherwise destruction will occur.

Why are pulley sizes important?

If the pulley size is too small, the alternator may over-speed and if it is too large the proper cut-in speeds may be wrong. For example an alternator is rated with a peak output at 2,300 revs/min. At a typical engine speed of 900 revs/min and a minimum required alternator speed of 2,300, a pulley ratio of approximately 2.5:1 is required. The maximum speed in this case has a 10,000-rev/min rating. Maximum engine speed is 2,300 in this case, so 2,300 multiplied by 2.5 = 4,000 revs/min. This falls well within operating speeds limits and is acceptable. A pulley giving that ratio would suit the service required.

Drive Pulley Selection Table

Engine Pulley	Pulley Ratio	Engine RPM	Alternator RPM
5 inch	2:1	2000	4000
6 inch	2.4:1	1660	4000
7 inch	2.8:1	1430	4000
8 inch	3.2:1	1250	4000

About overvoltage and surge protection

Zener diodes, which are often used, limit any potentially damaging high voltage spikes or peaks below a safe value, which otherwise could damage the regulator. The typical limiting voltages of Zener diodes in use are 25-30 volts for 14-volt alternators and 50-55 volts for 28-volt alternators.

How can interference be suppressed?

Alternator diode bridges create Radio Frequency Interference (RFI) or noise that can be heard on communications or electronics equipment. As a standard, install a 1.0-microfarad suppressor. In some cases, a suppressor is required in the main output cable. The NewMar 80-A and 150-A are designed for installation in the alternator output lead adjacent to the alternator. They will attenuate noise in the 70kHz to 100 MHz range that commonly affects GPS and communications radios.

Alternator Interference and Suppression

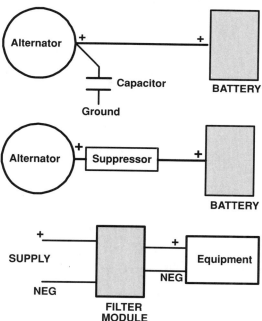

About filter units

The Adverc Copfilter has a passive input filter and transient suppressor, which is used to attenuate any voltage spikes caused by other equipment on the electrical system. A Low Loss regulator then maintains the output voltage at 11-15 volts. Where the input voltage is below the set output voltage, the output will equal the input voltage less 0.2 Volts. The unit will cut out if the temperature increases above 70°C, but will automatically reset when the unit has cooled. If the output connections are accidentally shorted or the input connections wrongly connected, the 10A fuse will blow.

Voltage Stabilization

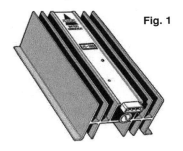

Fig. 1

Conditioning Module

Courtesy Adverc BM

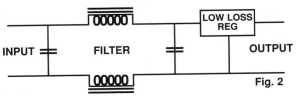

About alternator faults and failures

Failures in alternators are primarily due to the following causes, many of which are preventable with routine maintenance.

- **Diode Rectifier Bridge Failures.** Diode failures are generally attributable to the following causes along with simple overheating and the inability for the heat sinks to dissipate properly:

    Reverse Polarity Connection. This is a common occurence. Reversal of the positive and negative leads will destroy the diodes.

    Short Circuiting Positive and Negative. A short circuit will cause excess current to be drawn through the diodes and the subsequent failure of one or more of diodes; the most common cause here is reversing the battery connections.

    Surge. This occurs if the charge circuit is interrupted, most commonly when an electrical changeover switch is accidentally opened. A high voltage surge is generated due to the inductive effect of the field and stator windings.

    Spikes. This is a short duration high voltage transient. Inductive loads starting up, such as pumps cause spikes. Voltages several times greater than the nominal voltage can appear.

- **Winding Failures.** Stator winding failures result usually from overheating. This is normally due to insufficient ventilation at long term high outputs, causing insulation failure and intercoil short circuits.

- **Brush Gear Failures.** Brush gear failures are not that common in a properly maintained alternator but failures are generally due to worn brushes and sparking, and the symptoms are fluctuating outputs, and radio interference.

- **Bearing Failure.** The first bearing to fail is normally the front pulley bearing due to heat from belt and pulley and mechanical loads. Rotation by hand will usually indicate grating or noise.

What do the alternator terminal markings mean?

Alternators have a variety of different terminal markings. The main positive output of an alternator is usually marked as B+, BAT, BAT+, POS or just +. The main negative output is usually marked as D-, B-, GND, BAT-, E, D- or just -. The field connection is usually marked as DF or F. The auxiliary output for the warning light is usually marked as D+, 61, IND, L or AUX.

How does an alternator regulator work?

The function of the regulator is to control the output of the alternator, and prevent the output from rising above a nominal set level, typically 14 volts, which would otherwise damage the battery, alternator and other electrical equipment. The regulator is a closed loop controller, constantly monitoring the alternator output voltage and varying the field current in response to output variations. An alternator produces electricity by the rotation of a coil through a magnetic field and varying the level of the field current controls the output. This is achieved by varying that field current from the regulator, which goes through one brush and slipring to the rotor winding, and completing the circuit back through the other slipring and brush.

What is counter voltage?

During charging a phenomenon called "counter voltage" occurs. This is caused by the inability of the electrolyte in the battery cells to percolate at a sufficiently high rate into the plate material pores and subsequently convert both plate material and electrolyte. This causes the plate surface voltage to rise. The battery will resist charging and trick the regulator by indicating an artificially high voltage with the recognizable premature reduction in charging. It is often referred to as "surface charge".

What does regulator sensing mean?

Some alternators are machine sensed. This means that the regulator monitors the output terminal voltage and adjusts the alternator output voltage to the nominal value, which is typically 14 volts. The machine-sensed regulator makes no compensation for charging circuit voltage drops. Voltage drops include under-rated terminals, cables and the negative path back through the engine block. The battery-sensed unit monitors the voltage at the battery terminals and adjusts the alternator output voltage to the nominal voltage. The battery-sensed regulator compensates for voltage drops across diodes and charge circuit cables. The regulator by sensing the battery terminal voltage varies the output from the alternator until the correct voltage is monitored at the battery. Most smart regulators are battery sensed.

Regulator Sensing

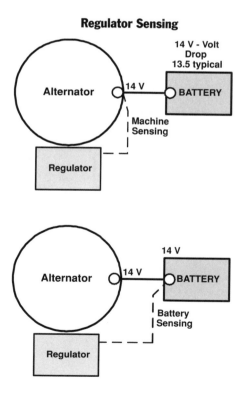

What does regulator polarity mean?

Regulators and field windings have two possible field polarities, and this does not have anything to do with the positive output of the alternator. The positive polarity regulator controls a positive excitation voltage. Inside the alternator, one end of the field is connected to the negative polarity. The negative polarity regulator controls a negative excitation voltage. Inside the alternator one end of the field is connected to the positive polarity.

Field regulator connection

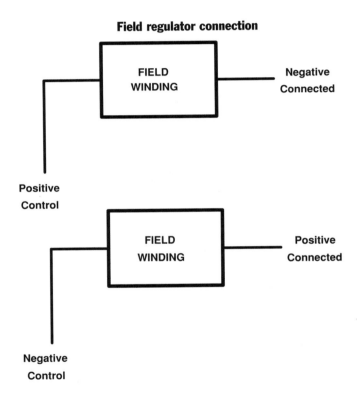

How does the Alpha 3-Stage charge regulator work?

This is a 3-stage regulator that has user definable settings for accept, float and time. The first step is the bulk charge phase, where voltage rises steadily up to approximately 14.2-14.4 volts, and maximum current output occurs up until approximately 80% charge level. The second step is the absorption or acceptance phase where the voltage is maintained constant and the current slowly reduces. The third step is the float phase where voltage reduces to approximately 13.8 volts and maintains a float charge to the battery.

Alpha Regulator
Courtesy Mastervolt

How does the Adverc controller operate?

This is an alternator controller and not a boost charge device, and it does not use the 3-stage charge concept. A cycle program is the basis of the charging system which applies charging voltages below and above the battery gassing voltage without actually causing gas generation. This is to recharge quickly without causing damage to battery or alternator. The regulator is designed for parallel connection to the existing regulator, giving some redundancy. Temperature compensation also takes place. There is a light warning system, with indication given for high voltage conditions or a loss of sense leads. I have an Adverc installed on my boat.

The cycle period. The cycle period consists of four 20-minute intervals followed by a 40-minute rest period. Voltage levels used within the charging cycle are a normal charge rate of 14.0 volts, and a high level of 14.4 volts to suit gel batteries without needing to change voltage settings.

Adverc regulator
Courtesy Adverc BM

How does the Balmar Max
Charge MC-612 Regulator operate?

This is a microprocessor-controlled unit with several user selectable multi-voltage variable-charge time programs for six battery types. The settings are made via dipswitches. The principle is the use of an automatic absorption time program, alarm outputs, LED status and alarm indicators. The amp manager function has a remote controlled power reducer if required. The unit has a data output port, and the option of a soft start and ramp up function, to save belt wear. There is an optional battery compensation sensor and alternator temperature sensor for over-temp protection.

How does the Balmar Max
Charge MC-412 Regulator operate?

This is a microprocessor-controlled unit with several user selectable multi-voltage variable-charge time programs for four battery types. It also has an LED display for program mode indication, and self-diagnostics. It has connections for warning light and electrical tachometer output. The amp manager function is included along with a data port. There is an optional battery compensation sensor and alternator temperature sensor for over-temp protection.

About the BRS-2 Regulator

This single stage regulator has a nominal setting of 13.7 volts, and has a range of settings for various battery types. It has connections for warning light and electrical tachometer output.

About the ARS-4 Regulator

This is a 3-step controller with user settings for deep cycle, gel, AGM and Optima batteries. It uses bulk, absorption and float charge principle. It also has an LED display for program mode indication, and self-diagnostics. It has connections for warning light and electrical tachometer output.

How does the PowerTap
Smart Regulator SAR-V3 operate?

The regulator uses a microprocessor controlled cycle type program. It has no operator adjustable functions with respect to the charging cycle, and operates based on 12 programmed charging cycles. Battery temperature compensation is incorporated and it is for use with P type alternators only. An alarm function uses a coded flash system. An over-voltage runaway circuit detects over-voltage conditions that occur when the regulator output has a short circuit and runaway. This is indicated via the alarm lamp circuit. Current limiting is via a user adjustable function that requires connection of an externally operated switch. The switch will reduce output to a relatively low level to avoid overheating of an alternator or to remove load off a smaller engine. The equalization function is a user adjustable feature that requires connection of an externally operated switch. The function enables an equalization current to be applied until battery voltage reaches 16.2 volts. The regulator has some very commendable features. The field output driver is short protected to prevent damage to the regulator in the event of a field circuit failure. Additionally all inputs are voltage transient protected, although normal precautions should still be installed. The lamp circuit is also overrated to provide alarm buzzer load capability. In addition, a voltage limit function enables charge voltage to be held at 13.8 volts to prevent halogen light damage during long night-motoring passages.

How does the PowerTap 3-Step Deep-Cycle Regulator work?

The 3-step device uses a step type program that is fully automatic and operates based on the charging cycles of absorption and float. The unit consists of a timer circuit rather than an intelligent program chip, and has simple battery and ignition inputs. Users are able to manually alter the absorption and float voltage settings, which is useful in applications such as NiCad cells that require different charging voltage levels. The manufacturer states that due to full alternator output requirement in step 1, many alternators may not be able to cope, and may suffer failure. This is generally due to windings overheating and diode failure. The regulator is suitable for P type alternators only (i.e. Bosch, Prestolite, Motorola, Valeo/Paris-Rhone etc). The regulator has the following control steps:

- The alternator is controlled to give full output until the absorption set point is reached. The time required to reach this level depends on the initial battery level and output speed of the alternator.

- The absorption set point is maintained for a period of 45 minutes (14.5 volts).

- The charge level reduces to the float voltage set point (13.8 volts).

How does the Next Step Regulator function?

The Next Step deep cycle regulator is an improved version of the 3-step unit. The unit is a microprocessor-controlled unit and incorporates temperature compensation. With full alternator output requirement in step 1, many alternators may not be able to cope, and may suffer failure. This is generally due to windings overheating and diode failure. Users are able to manually alter both absorption voltage and time as well as float voltage settings. The regulator has the following control steps:

- The alternator is controlled to give full output until the absorption set point is reached. The time required to reach this level depends on the initial battery level and output speed of the alternator.

- The absorption set point is maintained for a period of 45 minutes (14.5 volts).

- The charge level reduces to the float voltage set point (13.8 volts).

How does the Ideal Regulator Work?

Current is a factor in the charging process, and not just a voltage dependent device. This is not seen in many other regulator types. The regulator is used in conjunction with a digital circuit monitor. The regulator has the following control steps.

- **Delay Period.** A 20-second delay period after voltage is applied from ignition to allow engine speed to rise to normal running speed.

- **Ramping Up Period.** This is a controlled increase of alternator output over a 10-second period until default current limiting value is reached. It sensibly reduces shock loadings and allows belts to warm up, and reduces power line surges that occur when full outputs are applied.

- **Charge Cycle.** The charge cycle allows full alternator output with the battery voltage rising until the charged voltage is reached (14.3 volts).

- **Acceptance Cycle.** Charging continues at 14.3 volts until charge current decreases to a default value of 2% of capacity. Once the 2% level is reached, the acceptance hold cycle begins.

- **Acceptance Hold Cycle.** Charging is held at 14.3 volts and the charging current is monitored and continues for a minimum of 10 minutes. A maximum of 20 minutes is imposed on this cycle.

- **Float Ramp Cycle.** This is a transition phase between charged and float cycles. Voltage reduces to the float setting of 13.3 volts during the cycle.

- **Float Cycle.** The voltage is held constant at 13.3 volts.

- **Condition Cycle.** This is a manually activated function. Current is held at 4% of battery capacity, until a maximum voltage of 16 volts is attained. Once voltage reaches 16 volts, it is maintained until charge current falls to charge current % setting. The cycle then automatically terminates. On termination it reverts to the float ramp cycle to bring the voltage down.

What charge settings are required in different battery types?

Different battery types require different charge voltages and regulator systems should be adjusted to suit the installed batteries. Check with your battery supplier

Battery Regulator Charge Levels

Temp	FloodedHi/Float	Gel Hi/Float	AGM Hi/Float
90°F (32°C)	14/13.1	14/13.1	14.4/13.9
80°F (27°C)	14.3/13.3	14.0/13.7	14.5/14.0
70°F (20°C)	14.4/13.5	14.1/13.8	14.6/14.1
60°F (15°C)	14.6/13.7	14.3/13.9	14.7/14.2
50°F (10°C)	14.8/13.9	14.2/14.0	14.8/14.3

9. CHARGING CIRCUIT SYSTEM

The three-position changeover switch

The charging system on most engines uses the same cabling as the engine starter circuit, which consists of a switch with three positions and off. The center position parallels both battery banks. It is common to see both batteries left accidentally in parallel under load with the flattening of both. Many are quite unreliable and are subject to high resistances on contacts with large volt drops under high load conditions often causing starting and charging problems.

Single Engine Switch Charging Configuration

What is the relay or solenoid configuration?

This system enables separation of the charging system from starting circuits. The relay or solenoid does offer a point of failure if incorrectly rated for the job. The relay interconnects both batteries during charging, and separates them when off. This prevents discharge between the batteries. The relay-operating coil is interlocked with the ignition and energizes when the key is turned on. When modifying the system, it is necessary to separate the charging cable from the alternator to starter motor main terminal where it is usually connected. Relay ratings should at least match the maximum rated output of the alternator.

Autoswitch

How does a diode isolator charging system work?

A diode is a semiconductor one-way valve, consisting of an anode and cathode. It allows electrons to flow one way only and has a high resistance the other way. A diode isolator consists of two diodes with their inputs connected. They allow voltage to pass one way only, so that each battery has an output. This prevents any back feeding between the batteries. They are mounted on heat sinks specifically designed for the maximum current carrying capacity and maximum heat dissipation.

Single Engine Diode Charging Configuration

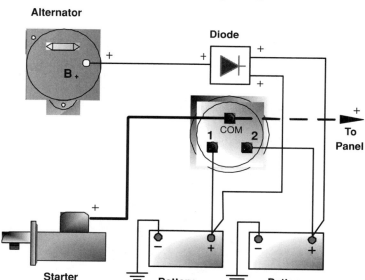

What about diode voltage drop?

A diode has an inherent voltage drop of typically 0.7 to 0.8 volts. This is totally unacceptable in a normal charging circuit. If the alternator is machine sensed and does not have any provision for increasing the output in compensation, the diode should not be used.

About diode selection

The diode isolators must be rated for at least the maximum rating of the alternator, and if mounted in the engine compartment must be overrated to compensate for the de-rating effect caused by engine heat.

About diode installation

Heat sink units should have the cooling fins in the vertical position to ensure maximum convection and cooling. Do not install switches in the cables from each output of the diode to the batteries.

Diode Isolators
Courtesy NewMar

How to test diode isolators

With engine running the diode output terminal voltages should be identical, and should read approximately 0.75 volt higher if a non-battery sensed regulator is being used. The input terminal from the alternator should be zero when the engine is off. Test with power off and batteries disconnected.

- Set the meter scale to ohms x1, and connect red positive probe to input terminal. Connect black negative probe to output terminals 1 or 2.

- If good the meter will indicate minimal or no resistance.

- Reverse the probes, and repeat the test. The reading should indicate high resistance, or over range.

How are two engines and two alternators set up?

Either in dual engine boats or in some boats there are two alternators installed to provide redundancy or improved charging capability. One alternator should charge the starting batteries and the second alternator should charge the house batteries. The alternators should not be charging in parallel to the same battery bank.

Dual Engine Diode Charging

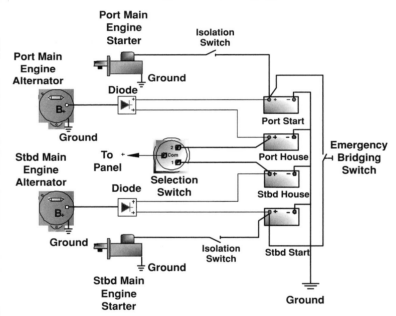

How are two engines and changeover switch set up?

In some boats with two engines and two alternators two changeover switches are used. The two circuits are generally wired up as shown in the circuit diagram.

Dual Engine Changeover Switch Charging

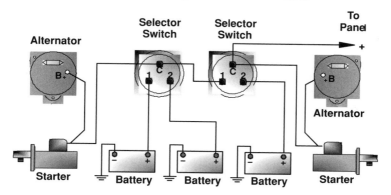

What are electronic battery switches?

These are also known as charge distributors or integrators, or combiners such as the NewMar Battery Bank Integrator (BBI). When a charge voltage is detected that exceeds 13.3 VDC, the unit switches on. The unit consists of a low contact resistance relay that closes to parallel the batteries for charging. When charging ceases and voltage falls to 12.7 VDC, the relay opens isolating the batteries. Other devices include the PathMaker which allow charging of two or three batteries from one alternator or battery charger; the Isolator Eliminator from Ample Power, which is a multi-step regulator that controls charge to the second battery bank, typically used for engine starting; and the Battery Mate from Mastervolt, which is a charge splitter that can supply three batteries, without voltage drop.

Electronic Switch
Courtesy Mastervolt

Why are the charging levels down?

The three most common causes are the drive belt is loose and slipping, there is oil on the belt, or there is a loose alternator connection. The next most common faults are a partial alternator diode failure, or a suppressor breaking down. Less common are diode isolator faults, negative connection faults and solder connection faults. Underrated cables are a cause where an alternator or battery bank capacity has been up-rated, or in a new installation. Less common are in-line ammeter and ammeter shunt faults.

Why are the batteries overcharging?

The most common cause is a regulator fault, or in some cases a regulator sense wire has come loose or broken or fallen off.

Why is there no charging?

The three most common causes are that the drive belt is loose or broken, there is an alternator diode bridge failure, or a regulator failure. A warning lamp failure or auxiliary diode failure can cause a problem. Less common are jammed brushes, stator and rotor winding failures. Although not a major cause the output and negative connections may be off.

Why is the alternator not working?

This initially depends on the lamp and the regulator. Using a voltmeter, check that the output across the main B+ terminal and negative or case increases to approximately 14 volts. No output indicates either total failure of alternator or regulator. Partial output indicates some diodes failed or a regulator fault. Often this is caused by operating a changeover switch.

What happens if the changeover switch is turned off under load?

If a changeover switch is operated under load, the surge will probably destroy the alternator diodes. Most switches incorporate an auxiliary make before break contact for connection of field. This advanced field switching disconnects the field and therefore de-energizes the alternator fractionally before opening of the main circuit. In modern alternators this is not used as most alternators have integral regulators and it is difficult to connect the switch into the field circuit.

Why is the charging ammeter fluctuating?

Usually the alternator brushes are sticking or there is a regulator fault. Another common cause is loose cable connections on the ammeter or at the alternator and battery end of the charging circuit.

Ammeter
Courtesy Blue Sea Systems

Adverc Monitor
Courtesy Adverc BM

10. BATTERY CHARGERS

How do AC powered battery chargers work?

Battery chargers are generally used as a primary charging source in boats that are alongside continuously or have AC generators on board. The AC mains voltage, either 230 or 110 volts AC, is applied to a transformer. The transformer steps down the voltage to a low level, typically around 15/30 volts depending on the output level. A full wave bridge rectifier similar to that in an alternator rectifies the low-level AC voltage. The rectifier outputs a voltage of around 13.8/27.6 volts, which is the normal float voltage level. Many basic chargers do not have output regulation. Chargers that do have regulation are normally those using control systems to control output voltage levels. These sensing circuits automatically limit charge voltages to nominal levels and reduce to float values when the predetermined full charge condition is reached.

Charger
Courtesy Mastervolt

What is a constant voltage charger?

It may also be called a constant potential charger if it operates at a fixed output voltage, typically 13.4-13.8 volts. It uses an SCR (Silicon Controlled Rectifier). The charge current decreases as the battery voltage reaches the preset charging voltage. Batteries can sustain damage if unsupervised as electrolytes evaporate and gas formation can be excessive. Additionally such chargers are susceptible to mains input voltage variations. If left unattended, the voltage setting must be below 13.5 volts, or batteries will be ruined through overcharging.

How does a ferro-resonant charger work?

These chargers use a ferro-resonant transformer, which has two secondary windings. One of the windings is connected to a capacitor, and they resonate at a specific frequency. Variations in the input voltage cause an imbalance, and the transformer corrects this to maintain a stable output. These chargers have a tapered charge characteristic. As the battery terminal voltage rises, the charge current decreases. Control of these chargers is usually through a sensing circuit that switches the charger off when the nominal voltage level is reached, typically around 15% to 20% of charger nominal rating.

How does a switch-mode type charger work?

Compact switch-mode chargers are becoming increasingly popular due to their compact size and very low weights. These charger types convert the input line frequency from 50 to 150,000 hertz, which reduces the size of transformers and chokes used in conventional chargers. An advantage of these chargers is that line input and output are effectively isolated eliminating the effects of surges and spikes. These chargers are my personal choice and I have one installed. The chargers are battery-sensed, temperature-compensated, have integral digital voltmeters and ammeters, and are physically very compact. They also are smart types that have selectable charge modes including bulk, acceptance and float and selectable voltages to suit gel, AGM or flooded cell batteries.

Battery charger installation

Chargers should be mounted in a dry and well-ventilated area. Always switch off battery charger during engine starting if connected to the starting battery. The large start load can overload the charger. Proper bolt on terminal lugs should be used on cables if the charger is permanently installed, not clips. Switch off the charger before connecting or disconnecting cables from battery, as sparks may ignite gases. Do not operate a large inverter off a battery with a charger still operating. The large load can overload the charger and may cause damage to circuitry.

How are multiple battery banks charged?

Most marina-based boats have a charger connected permanently to charge a single house battery bank although many boats have multiple house banks and twin engines with separate batteries. A separate battery charger is required or a method of splitting the charge to each battery. The two or more batteries under charge should also have bridged negatives if the two systems are electrically isolated. Remember that gel cells or AGM batteries may have different requirements and this should be checked prior to using any system, as batteries when fully charged can loose water rapidly if charging is imprecise.

Multiple output chargers

Install a battery charger with multiple outputs, such as those from NewMar or StatPower, where each battery bank has its own isolated charging outputs. This prevents any interaction and is an efficient way of having two or more separate chargers.

Using diodes

A diode isolator can be used to split the charge between the two or three battery banks. For three battery banks, use 2 diode isolators, and link the diode isolator inputs. There are problems of voltage drop across the diode that have to be considered, and battery chargers with battery sensing are required to compensate for this. The typical voltage drop is around 0.7 volt so the charger outputs without sensing will require adjustment of output voltage up an additional level equivalent to the drop.

Using a relay or solenoid

A relay or solenoid can be used to direct the charge current to each battery bank. This is activated either with the monitored charging voltage or via a manually operated switch. The configuration effectively parallels all of the batteries to form a single battery bank. Possible systems include the NewMar Battery Bank Integrator (BBI). When a charge voltage is detected that exceeds 13.3 VDC the unit switches on. The unit consists of a low contact resistance relay that closes to parallel the batteries for charging. When charging ceases and the voltage falls to 12.7 VDC, the relay opens isolating the batteries. Another similar device is the PathMaker from Heart Interface. These devices allow charging of two or three batteries from one alternator or battery charger. The units use a high current switch rated at 800 and 1600 amps for alternator and charger ratings up to 250 amps.

Using Smart Devices

These are intelligent charge distribution devices. The Ample Power Isolator Eliminator is a multi-step regulator that controls charge to the second battery bank, typically the one used for engine starting. It is temperature compensated like an alternator control system and is effectively a secondary charger. The AutoSwitch from Ample Power, which is a smart solenoid system, is another system. An electronic sensing circuit will enable the setting of the different modes. One mode is a timed function that terminates the charging to the second paralleled start battery once the period expires. There is also a voltage mode, which disconnects the second battery after the preset voltage is reached. Other devices such as Charge-Link and Echo-charge perform a similar function. These smart devices reduce the chances of overcharging secondary batteries such as the start or generator battery.

Charge Distributor
Courtesy NewMar

11. ALTERNATIVE CHARGING, SOLAR AND WIND

What about DC charging systems?

An alternative or addition to main propulsion energy charging systems is a dedicated engine powering an alternator. Balmar in the United States has a unit driven by a FW cooled 13-hp Yanmar diesel. It is installed with a 310-amp brushless alternator, Max Charge regulator, and small start battery alternator. The Ample Power Genie uses a seawater cooled Kubota diesel fitted with a 120-amp alternator and Smart regulator system. A new system known as the WhisperGen, based on a principle developed in 1816, uses a Sterling cycle machine. This consists of a continuous combustion process, noiseless as the motor/generator is hermetically sealed. It requires no oil lubrication as there are no moving parts, and there are no exhaust fumes. The DC output is from a permanent magnet DC generator and ratings are in the 4-6 kW range.

About solar charging principles

The fundamental process of a solar cell is that when light falls on to a thin slice of silicon P and N substrate, a voltage is generated. This is called the photovoltaic principle. Cells consist of two layers, one positive, and one negative. When light energy photons enter the cell, the silicon atoms absorb some photons. This frees electrons in the negative layer, which then flow through the external circuit (the battery) and back to the positive layer. When manufactured, the cells are electronically matched and connected into an array by connecting in series to form complete solar panels with typical peak power outputs of 16 volts. There are a number of solar cell types, based on the cell material or structure used.

What does mono-crystalline mean?

Pure, defect-free silicon slices from a single grown crystal are used for these structures. The cell atomic structure is rigid and ordered and unlike amorphous cells cannot be easily bent. The cells are approximately 12%-15% efficient. The thin pure silicon wafers are etched within a caustic solution to create a textured surface. This textured surface consists of millions of four sided pyramids, which act as efficient light traps reducing reflection losses. Panels are made by interconnecting 34-36 wafers onto a glass back and encapsulated.

What does polycrystalline mean?

These cell types use high purity silicon 0.2mm wafers from a single block, and are high power output cells. The wafers are bonded to an aluminium substrate, and Solarex cells are covered with a tempered iron glass, and a titanium dioxide anti-reflective coating to improve light absorption. The polycrystalline cell has better low light angle output levels and is now the most commonly used.

What does amorphous silicon mean?

These cells are formed from several layers applied to a substrate. They have a characteristic black appearance. Solarex cells use a tin oxide coating to improve conductivity and light absorption. Unlike crystalline cells, these thin film panels have a loosely arranged atomic structure and are much less efficient. They do have the advantage that the cells can be applied to flexible plastic surfaces and as such flexible panels are made. Additionally they are capable of generating under low light conditions.

About solar ratings and efficiency

Efficiency is at an optimum when a solar panel is angled directly towards the sun and manufacturers rate panels at specific test standards. Output ratings are normally quoted to a standard, typically 1000W/m^2 at 77°F (25°C) cell temperature, and the level of irradiance is measured in watts per square meter. The irradiance value is multiplied by time duration to give watt-hours per square meter per day. Location and seasonal factors affect the amount of energy available. Cells are approximately 15% efficient and start producing a voltage as low as 5% of full sunlight value. Solar angles are important to the efficiency of panels. With the sun at 90° overhead, panels give 100% output. When angled at 75°, the output falls to approximately 95%, at 50° output falls to 75%, and a lower light angle of 30° gives a reduction to 50%. Many panels now will give some output on dull days.

How can the solar system affect the alternator charging?

There is often an interaction between solar panels and alternator charging regulators, when they are left on during engine charging periods. In many installations, solar panels are not regulated, and it is quite common to see a voltage of up to 16 volts or more across the battery. This can damage the batteries. When an alternator regulator senses this high voltage level, it artificially "sees" this as a fully charged battery, and as a result, the alternator does not charge the battery, or at a very low charge rate. When installing panels and regulators, an isolation switch should be installed in the incoming line to the panel so that it can be switched out of circuit. Alternatively, the solar panel can be automatically disconnected via a relay so that the solar panel output does not impress a higher voltage and confuse the alternator regulator.

How are solar panels regulated?

In any panel greater than a small 12-15 watt unit, a shunt regulator is required to restrict the voltage to a safe level. It is not uncommon to have voltage levels rise to 15-16 volts, which can boil the battery dry over any extended and unsupervised period. The external regulator functions simply to limit panel output voltage to a safe level and prevent damage to a battery. Units may be simple and limit voltage to 13.8 volts, the maximum float level, dissipating heat through a heat sink. More sophisticated units incorporate an automatic boost level of 14.2 volts and a float setting of 13.8 volts. The regulator will float charge the battery until a lower limit of approximately 12.5 volts is reached before switching to boost charge. The regulator eliminates the need for an additional blocking diode.

Solar Panel Regulator
Courtesy Blue Sea Systems

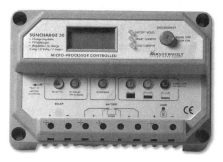

About solar panel diodes

Most panels have diodes installed. While a diode can reduce the voltage by approximately 0.75V, if you are installing a couple of three amp panels, which is typical, you will require a regulator to reduce the voltage to avoid overcharging and damaging your batteries. If the regulator is a good unit the control will float between 14.5 and 13.8 volts, so this voltage drop will not be a major problem. If the regulator has reverse current protection such as a diode, then the panel-installed unit can be removed to increase the input voltage to the regulator, which will give a marinally higher output. If you are not going to regulate the solar supply, failure to install or leave the diode installed may result in a flat battery overnight. There are two diode purposes. By-pass diodes are normally factory installed within solar module junction boxes. By-pass diodes are used to reduce power losses that might occur if a module within the array is partially shaded. For 12-volt systems, these offer sufficient circuit protection without the use of a blocking diode. In 24-volt systems with 2 or more modules connected in series, the solar modules should be connected in individual series circuits. To obtain the required total array current the circuits should then be paralleled. If one module of a parallel array is shaded, reverse current flow may occur. Blocking diodes are often connected in series with the solar panel output to prevent discharge of the battery back to the array at night, but not all manufacturers install them as standard. Most solar regulators will often have the diode incorporated.

How to install solar panels

Cabling should be properly rated to avoid voltage drop. As cable is external, use tinned copper marine cable. Most panels have weatherproof connection boxes and connections should be simply twisted and terminated in terminals. Do not use connectors or solder the wire ends. Manufacturers also specify grounding of array or module metallic frames. Always cover solar panels to prevent a voltage being generated during installation or removal so that accidental short-circuiting of terminals or cables cannot occur. Each panel should be securely mounted and able to withstand mechanical loads. Ideally, they should be oriented to provide unrestricted sunlight from 9 to 3 p.m. solar time. Allow sufficient ventilation under the panel. Most panels in frames have sufficient clearance incorporated into them. Excessive heat levels will reduce output and damage cells.

About solar panel maintenance

Maintenance requirements of solar panels are minimal. Panels should be cleaned periodically to remove salt deposits, dirt and seagull droppings. Use water and a soft cloth or sponge. Mild, non-abrasive cleaners may be used, do not use any scouring powders or similar materials. Make sure the terminal box connections are secure and dry. Fill the box with silicon compound.

Troubleshooting solar panels

Faults are normally the results of catastrophic mechanical damage when people walk on them or some dropped object impacts on the surface. A single cell failure will not seriously reduce performance as multiple cell interconnections provide some redundancy. Reliability is very high and manufacturers give 10-year warranties to support this. Faults can be virtually eliminated by proper mounting and regular maintenance. As with all electrical systems, the most common faults are cable connections. Panels are de-rated for several conditions, such as high temperatures, a panel operates 68-77°F (20-25C°) higher than the ambient temperature. Check the regulator output for the correct rated voltage, typically it is 13.2 volts. Check the regulator input, the voltage will be typically 14 volts or greater at full panel output. Disconnected from the battery it can have an open circuit voltage up to or exceeding 17 to 18 volts. Check the panel junction boxes for moisture or corroded connections. Check that panels are clean and undamaged and for panel shadowing by spars that affect output. Make sure panel azimuth and tilt angles are best for solar collection.

Solar Panels
Courtesy Mastervolt

About wind charging output ratings

Wind generators appear more effective in some areas than others. In the Caribbean they are very effective where good trade winds blow. In the Mediterranean, solar power is considered more efficient. If your cruising lifestyle also takes you primarily to sheltered anchorages, they may not be a practical proposition. It is however at anchorages where wind generators are the most useful and can give 24-hour charging. The average wind generator typically produces anything from 1 amp to 15 amps maximum depending on the wind speed. Ratings curves are always a function of wind speed and are quoted at rated output voltages. Check each model to suit your particular requirements.

Rutland Wind Generator Circuit

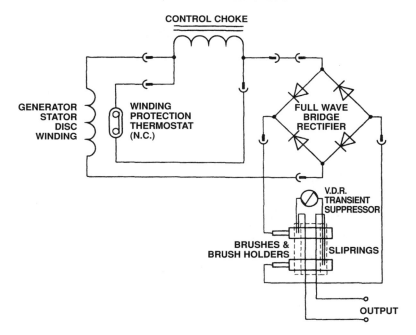

Wind generator types explained

Essentially a wind generator is either a DC generator or alternator driven by a propeller. In the United States the trend is normally for large two- or three-bladed DC generator units. The UK/European trend is for smaller diameter six-bladed AC alternator units. These units incorporate a heavy hub that acts as a flywheel to maintain blade inertia. Many units have a permanent magnet rotor, with up to 12 poles. A three-phase alternating current is generated and rectified to DC similar to engine driven alternators. The aerodynamically shaped Air-X Marine is a three-bladed unit and has a brushless permanent magnet alternator with internal regulator. My own observations show the six-bladed units to be the most efficient in relatively sheltered locations with lower wind speeds, and three-bladed in voyaging and higher wind locations.

What about charge regulators?

A regulator is required to both limit normal charging voltages to a safe level (14.5 volts) and to limit output at high wind speeds. Normally a shunt regulator is preferred over a normal solar panel regulator, as it is more suited to constant loads. Shunt regulators divert excess current to a resistor which functions as a heater and dissipates heat through a heat sink. If series regulators are used, a power zener diode should be installed to provide some load when the battery is fully charged. A 12-volt system should use an 18-volt zener diode. The zener must be rated for at least half rated generator output. Like solar panels installations, interaction may occur with alternator charging systems. The charging should be either switched out of circuit or diverted to a battery other than the sensed one (e.g. start battery). Some units incorporate a choke to limit the charge produced at high wind speeds. A number of generators incorporate a winding embedded thermostat which opens in overload conditions, and this operates when the winding overheats. Some units incorporate a transient suppressor, which is installed to minimize the effects of intermittent spikes being impressed on the charging system. These would damage the rectifier and onboard electronics. The suppressor is usually a Voltage Dependant Resistor (VDR). The Air-X Marine unit uses some microprocessor-based smart charging concepts. The charge controller will regularly cease charging, monitor the battery voltage, then compare it to the voltage setting. If the battery is charged, it will completely shut off all charge current to the battery. When the battery approaches full charge, this process is carried out more frequently. When the battery is fully charged, it will slow to virtual full stop condition. When the battery voltage falls lower than the voltage set point, charging will restart.

About wind generator installation

A common arrangement is a stern post, which keeps the blades clear of crew. One of the major complaints is that under load the wind generator creates vibration. It is essential that the post be as thick sectioned as possible and supported. Mountings can also be cushioned with rubber blocks or similar material to reduce the transmission of vibrations.

Troubleshooting wind generators

Always secure the turbine blades when installing, servicing or troubleshooting a wind generator. If no ammeter is installed on the main switchboard, install an ammeter in line and check the charging current level. If there is no output, check the system according to the manufacturer's instructions. If there is no output and the generator has brushes, make sure they are free to move and are not stuck. Many generators do not have brushes and commutators, but a set of sliprings are installed with brushes to transfer power from the rotating generator down through the post to the battery circuit. They can jam and on rare occasions cause loss of power. Some generators have a winding imbedded thermostat. Check with a multimeter that it is not permanently open circuited. If it is open circuited the generator will not charge. The thermostat will open in high wind charging conditions. If the thermostat has not closed after these conditions and the generator case is cold, the thermostat is defective. Regrettably there is nothing that can be done to repair it unless a new winding is installed. To get the generator back into service, connect a bridge across the thermostat terminals. Check the rectifier to ensure that it is not open or short-circuited. If the generator output is correct, check the regulator is not malfunctioning. The voltage input may be in the range 14-18 volts, and the output approximately 13-14 volts. Ensure all electrical connections are secure and in good condition.

What about vibration?

Bearing wear will cause excessive vibration. If the unit is a few years old, renew the bearings. Vibration can also be caused by damage to one or more blades, and these should be carefully examined for damage that may be causing imbalances. One cause of vibration is the designed bending or feathering of the blades. Air-X Marine controls the rotation when nearing rated output so as to reduce the noise that is heard in high wind gusts as blades bend causing the flutter. The output is cut back at peak output through speed control.

Acknowledgements

Thanks and appreciation go to the following companies for their assistance. Readers are encouraged to contact them for equipment advice and supply. Quality equipment is part of reliability!

Adverc	www.adverc.co.uk
Air-x Marine	www.windernergy.com
Ample Power	www.amplepower.com
Balmar	www.balmar.net
Blue Sea	www.bluesea.com
NewMar	www.newmar.com
Marlec	www.marlec.co.uk
Mastervolt	www.mastervolt.nl
Rolls	www.rolls.com
Surrette	www.surrette.com
Volvo	www.volvo.se
Xantrex	www.xantrex.com

The Marine Electrical School

This book contains the material for Module 101 of the Certificate in Basic Marine Electrical and Electronics. Log on to www.marineelectrics.org for course details.

Index

Other books by John C. Payne:

MOTORBOAT ELECTRICAL & ELECTRONICS MANUAL
by John C. Payne

Following the international success of *Marine Electrical and Electronics Bible*, Payne turns his talents from sailing boats to powerboats. This complete guide, which covers inboard engine boats of all ages, types, and sizes, is a must for all builders, owners, and operators. Payne has put together a concise, useful, and thoroughly practical guide, explaining in detail how to select, install, maintain, and troubleshoot all electrical and electronic systems on a boat.

Contents include: diesel engines, instrumentation and control, bow thrusters, stabilizers, A/C and refrigeration, water and sewage systems, batteries and charging, wiring systems, corrosion, AC power systems, generators, fishfinders and sonar, computers, charting and GPS, radar, autopilots, GMDSS, radio frequencies, and more.

"... tells the reader how to maintain or upgrade just about every type of inboard engine vessel." *Soundings*

MARINE ELECTRICAL & ELECTRONICS BIBLE
by John C. Payne

"Everything a sailor could possibly want to know about marine electronics is here...as a reference book on the subject it is outstanding."
Classic Boat

"A bible this really is...the clarity and attention to detail make this an ideal reference book that every professional and serious amateur fitter should have to hand." *Cruising*

"...this is, perhaps, the most easy-to-follow electrical reference to date."
Cruising World

"All in all, this book makes an essential reference manual for both the uninitiated and the expert." *Yachting Monthly*

"...a concise, useful, and thoroughly practical guide.... It's a 'must-have-on-board' book." *Sailing Inland & Offshore*

SHERIDAN HOUSE
America's Favorite Sailing Books
www.sheridanhouse.com